YOUR KNOWLEDGE HAS VALUE

- We will publish your bachelor's and
  master's thesis, essays and papers

- Your own eBook and book -
  sold worldwide in all relevant shops

- Earn money with each sale

Upload your text at www.GRIN.com
and publish for free

James Sutanto

Environmental Study of Lead Acid Batteries Technologies

GRIN Verlag

Bibliografische Information der Deutschen Nationalbibliothek:

Die Deutsche Bibliothek verzeichnet diese Publikation in der Deutschen National-
bibliografie; detaillierte bibliografische Daten sind im Internet über http://dnb.d-
nb.de/ abrufbar.

Imprint:

Copyright © 2011 GRIN Verlag GmbH
Druck und Bindung: Books on Demand GmbH, Norderstedt Germany
ISBN: 978-3-656-03384-4

This book at GRIN:

http://www.grin.com/en/e-book/179643/environmental-study-of-lead-acid-batteries-
technologies

Environmental Study of Lead Acid Batteries Technologies

James Sutanto

Electronic Science and Technology, Xi'an Jiao Tong Liverpool University, Suzhou,China

Abstract: *This article presents the results of lead acid battery usage in the late 2000s. In this study, the usage of the lead acid battery was increased every year. However, there were several limitations due to the lead acid battery such as, the health effect, cause explosion. On the other hand, Lead–acid battery recycling is one of the most successful recycling programs in the world, which going to be encouraged to every people, instead using disposable batteries.*

Introduction

At the end of the last Millennium, the term 'sustainability' became an overall guiding principle for human development. Its success stems from the underlying reflections on existential problems of mankind perceived at that time: increasing concern over exploitation of natural resources and economic development at the expense of environmental quality.

The word sustainability is defined from the Latin *sustinere* (*tenere*, to hold; *sus*, up). Dictionaries provide more than ten meanings for *sustain*, the main ones being to "maintain", "support", or "endure". However, since the 1980s *sustainability* has been used more in the sense of human sustainability on planet Earth and this has resulted in the most widely quoted definition of sustainability and sustainable development.

In the 20th century, life would not be possible without lead acid batteries. Therefore, lead acid battery are used so extensively all around the world. (Battery Council International, 2004) Batteries provide backup power for telecommunication companies, mobile phone, computer system and power two way radio. Moreover, batteries play a hidden but primary support role in nearly every kind if IT system. Unremitting power is becoming increasingly essential for the functioning of sensitive test utilities, computing devices, and networks. (Frost and Sullivan, 2004a).

Other than fluorescent light bulb, batteries are the most common items that contain high concentrations of poisonous materials. The majority use of lead for batteries are world wide apply. In general, the battery industry has created significant efforts to recycle and reduce poisonous materials (ICCL, 2001). In addition, all types of batteries are now globally classified as poisonous waste and disposal of batteries is regulated (California State University, Long Beach, 2005).

Environmental Concerns

The main environmental concerns related to batteries are:

- Health effects on industrialized workers
- Health effects on building occupants in a fire
- Lead mining
- Disposal at the end of useful life

Scale of the problem

In the U.S 350 million rechargeable batteries are sold every year and 3 billion batteries are thrown away (APC, 2005). In 2001, Californians purchased primary and secondary batteries over 500 million (Earth911, 2005 and Rechargeable Battery Recycling Corporation, 2003).

Batteries – The Basics

Batteries either directly power or support for desktop, laptops computers, telephone system, and any other electrical devices. The use of batteries is growing for three main reasons:1) The continuing proliferation of electronic gadgets that require an uninterrupted supply of power, 2) Increased in unreliability of commercial power sources, and 3) rapid growth in the use of mobile IT devices.

There are two general types of batteries.

Primary batteries- Alkaline, lithium, silver oxide, zinc-air, zinc-carbon and zinc-chloride are non-rechargeable batteries that are thrown away or recycled after discharged, and have been generally used mostly in consumer settings.

Secondary batteries- Nickel-cadmium, nickel metal hydride, lithium-ion and lead-acid batteries are rechargeable, and are more heavily used in business settings than primary batteries. In addition, Frost and Sullivan (2004b) stated that about 50 percent of the UPS market uses lead-acid technologies and more than 50% of SLA sales are for telecom appliances.

The main secondary battery categories are:

Stationary Lead-Acid (SLA)- The primary constituents of an SLA are lead (71-76%), sulfuric acid electrolyte (16-19%), and 8-10% other materials, including case, vnts and other components. Battery production remains America's main consumer of lead, consuming 88% lead used in the U.S. (USGS, 2005a).

Environmental Concerns about batteries

Health Effects on Manufacturing Workers

Lead is not readily absorbed through skin. On the other hand, lead dust or fumes may lead to eye and respiratory problems, as well as headaches, vomiting, abdominal pain, diarrhea, severe cramping, and difficulty in sleeping. Lead may also cause anemia, kidney and nervous system damage, as well as reproductive harm in both male and female. The International Agency for Research on Cancer (IARC) and the EPA categorize lead as a possible human carcinogen (Exide, 2003b).

In factories such as lead smelting and mining, as well as assembling settings where lead and lead-based items are present, workers are victim to exposure and poisoning. People who live near lead processing are at particular risk through factories emissions, industry waste disposal, and contaminating of soil or clothing and footwear from peoples working in those industries.

Research in the United Studies found that lead dust on parents work clothes, hair, skin, and vehicles or home activities that contaminated eating areas can infect the children which high levels of lead in their blood (Leadpoison.net, 2002). A report by Centers for Disease Control and Prevention agency NIOSH stated that take home contamination risks for family have been mostly ignored that problems are seems to continue and grow in the future (National Institute for Occupational Health and Safety, 2002).

Health Effects on Building Occupants in a Fire

Batteries cause clear exposures during the material manufacturing process and after disposal, but only potential risks during use. Under normal conditions, SLA batteries do not create lead dust, vapors or fumes. On the other hand, hazardous exposure may occur when SLAs are overheated, oxidized, or damaged to generate dust, vapor, or fumes such as during fires in buildings.

The major element of SLAs after lead is sulfuric acid. Under normal conditions, SLAs do not create acid vapors or haze. However such as lead, like hazards can be produced when fire occurs. Under such conditions, acid vapors or mist may cause severe stomach, skin, and respiratory irritation, as well as burns and ulceration.

Lead mining

Lead is used by all developed nations. The USA is by far the largest consumer, with some countries in Asia (China, Japan Korea) and Europe (France, Germany, Italy, and UK) also using large amounts. Most of the lead is used for batteries, an appliance which has grown greatly in importance. Approximately 3 million metric tons of lead ore is mined worldwide each year with 75% of the manufacture coming from six countries: Australia, Canada, China, Mexico, Peru, and U.S.A (ICCL, 2001 and ILZG, 2005).

The ecological effects of lead mining are numerous: the majority of lead is mined underground and near the surface (EERE, 2003) which often due to open mine shafts, collapse mine shaft, and subsidence areas which cause property damage and produce avenues for water to enter and leave the mines.

Disposal at the end of useful life

While the majority of lead acid batteries are appropriately disposed of and recycled in the U.S., the enormous majority of both primary and secondary batteries sold are not ending up in the waste stream.

<u>Lead-Acid Batteries</u>

50% of the worldwide lead used in producing is created from the recycling of existing lead products. Recycling percentages are higher in developed countries, mainly U.S. where 80% or more of domestic lead expenditure was recovered from used batteries (ICCL, 2001).

However, mainly most commercial and developed SLAs are likely to be recycled, others are not. Smaller UPS's for desktop, workgroup, or telecom closet encountered for 37% of total North American UPS delivery in 2002 (Venture Development Corporation, 2004). Various products are used in marketable applications as additional support power units in larger IT systems or in Small Office/Home Office (SoHo) applications. In this case, users are normally unaware or unconcerned that their small UPS contains a preserved lead-acid battery and that state and federal laws forbid removal of lead batteries as normal waste.

Hazards

Explosion

A battery explosion is caused by the misuse or malfunction of a battery, such as attempting to recharge a primary (non-rechargeable) battery, or short circuiting a battery. With car batteries, explosions are most likely to occur when a short circuit generates very large currents. In addition, car batteries liberate hydrogen when they are overcharged (because of electrolysis of the water in the electrolyte). Normally the amount of overcharging is very small, as is the amount of explosive gas developed, and the gas dissipates quickly. However, when "jumping" a car battery, the high current can cause the rapid release of large volumes of hydrogen, which can be ignited by a nearby spark (for example, when removing the jumper cables).

When a battery is recharged at an excessive rate, an explosive gas mixture of hydrogen and oxygen may be produced faster than it can escape from within the walls of the battery, leading to pressure build-up and the possibility of the battery case bursting. In extreme cases, the battery acid may spray violently from the casing of the battery and cause injury. Overcharging—that is, attempting to charge a battery beyond its electrical capacity—can also lead to a battery explosion, leakage, or irreversible damage to the battery. It may also cause damage to the charger or device in which the overcharged battery is later used. Additionally, disposing of a battery in fire may cause an explosion as steam builds up within the sealed case of the battery.

Leakage

Many battery chemicals are corrosive, poisonous, or both. If leakage occurs, either spontaneously or through accident, the chemicals released may be dangerous.

For example, disposable batteries often use zinc "can" as both a reactant and as the container to hold the other reagents. If this kind of battery is run all the way down, or if it is recharged after running down too far, the reagents can appear through the cardboard and plastic that forms the residue of the container. The active chemical leakage can then damage the equipment that the batteries were inserted into. For this reason, many electronic device manufacturers recommend removing the batteries from devices that will not be used for extended periods of time.

Recycling

Lead–acid battery recycling is one of the most successful recycling programs in the world. In the United States 97% of all battery lead was recycled between 1997 and 2001. An effective pollution control system is a necessity to prevent lead emission. Continuous improvement in battery recycling plants and furnace designs is required to keep pace with emission standards for lead smelters (Battery Council International, 2009).

According to Waste Age, lead-acid batteries have the highest recycling rate of any product sold in the U.S. since they are easily returned when a new battery is purchased and because a battery's lead and plastic components are valuable. More than 80% of the lead produced in the U.S. is used in lead-acid batteries.

A typical battery contains between 60% and 80% recycled lead and plastic and consists of the polypropylene casing; lead terminals and positive and negative internal plates; lead oxide; electrolyte, a dilute solution of sulfuric acid and water; and plastic separators that are made from a porous synthetic material.

Several recycling markets are utilized with discarded batteries. Polypropylene casings are processed and recycled into new battery casings. Lead is recycled into lead plates and other battery parts. Battery acid is either neutralized, treated and discharged into sewers, or processed into sodium sulfate, which is a powder used in laundry detergent, glass and textile manufacturing.

Approximately 1.88 million tons of batteries in municipal solid waste weight are recycled every year for a remarkable 96.9% recycling rate, aided by laws in 37 states requiring retailers to collect old lead-acid batteries from customers who buy new ones. Only 60,000 tons, or less than 0.1 % of discarded municipal solid waste, is land filled since 41 states ban the disposal of lead acid batteries

Fig.1 Approximately 81 percent of all lead consumed in the United States goes into lead-acid storage batteries. A typical lead-acid automotive battery contains about 9.75 kilograms of lead. Photograph courtesy of Delphi Automotive Systems, Inc. (http://pubs.usgs.gov/sir/2006/5155/sir20065155.pdf)

Environmental Benefits
Lead-acid batteries are the environmental success products. More than 97 percent of all battery lead is recycled. Compared to 55% of aluminum soft drink and beer cans, 45% of newspapers, 26% of glass bottles and 26% of tires, lead-acid batteries top the list of the most highly recycled consumer product.

The lead-acid battery gains the environmental edge from its closed-loop life cycle. The typical new lead-acid battery contains 60 to 80 percent recycled lead and plastic. When a spent battery is collected, product is sent to a permitted recycler where, under strict environmental regulations, the lead and plastic are reclaimed and sent to a new battery manufacturer. The recycling cycle goes on indefinitely. That means the lead and plastic in the lead-acid battery in consumer's car, truck, boat or motorcycle going to continue recycled in a large numbers. This makes lead-acid battery disposal extremely successful from both environmental and cost perspectives.

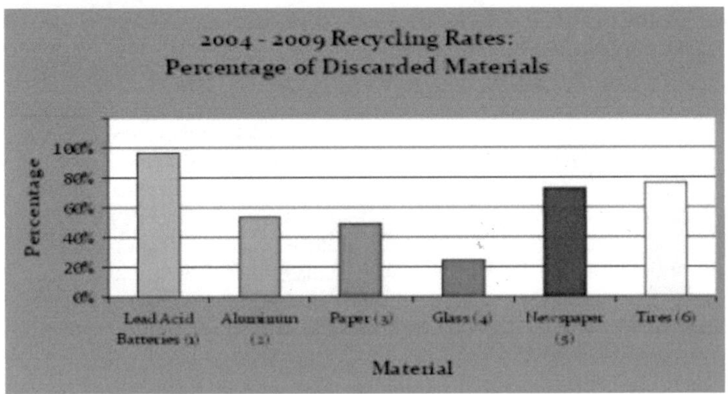

Fig. 2 Recycling Chart data for newspapers, glass bottles, tires and aluminum cans as compared to lead-acid batteries (http://www.batterycouncil.org/LeadAcidBatteries/BatteryRecycling/tabid/71/Default.aspx)

Discussion

Safe Handling

When handling batteries that still contain acid, appropriate Personal Protective Equipment (PPE) should be worn. This includes coveralls, protective glasses and gloves. This equipment can be purchased by specialist safety gear suppliers (refer to contacts list for suggested suppliers). When lead plates are melted to form ingots for efficient shipping, extreme care needs to be taken to prevent inhalation of toxic lead fumes. A well fitted respirator with a cartridge suitable for lead must be worn around the heater and downwind, in addition to coveralls, glasses and gloves (Lead Acid Battery Management, 2010).

Undesirable Practices
- Emptying of acid in batteries to the ground and waterways,
- Lead recovery at a domestic level to make fishing sinkers and weights for diving belts,
- Storage of batteries for any length of time while still connected
- Storage of batteries outside and uncovered.

Desirable Practices
- Collecting batteries for either recycling or proper landfill disposal,
- Reuse and recycling of as much of the battery as possible,
- Shipping batteries to secure recycling facilities for recycling,
- Storage of batteries in a safe storage facility as detailed below

Lead-acid car batteries are one area where the US is doing very well when it comes to battery recycling. Almost all of them are recycled, which is great news, because the lead in car batteries is very toxic and not something we want floating around in our environment.

Collection

Several different sectors of society need to be involved to ensure the collection system is effective, including scrap dealers, battery dealers and consumers. Collection centers should be located at retailers, service stations and other places where new batteries can be purchased. In this way disused batteries can be collected for forwarding to treatment. A collection point at landfill sites should also be considered. At the collection center the disused batteries need to be stored so as to minimize leakage. Ideally, storage in an acid-resistant container is preferred, although this is often not possible due to cost.

- Leaking batteries should be stored on a bunded pallet,
- The storage place should be sheltered from rain, water sources andvaway from heat,
- The ground of the storage place should have a ground cover, preferably plastic or any other acid-resistant material, that retains any leakage and directs it to a collection container for disposal,
- The storage place should have good ventilation to prevent hazardous gas accumulation,
- The storage place should have restricted access and be identified as a hazardous material storage place.

Any storage place should not accumulate large amounts of batteries (i.e. over 100), and must not be considered as a permanent storage facility. Limiting the quantity of batteries decreases the chances of environmental and workplace safety accidents. Collection points must not sell their batteries to unauthorized lead smelters. Unauthorized smelters are one of the biggest polluters of lead contamination, both to humans and the environment.

Conclusion

Batteries enable our mobility, so likely society will be using lots more batteries in the future. But to ensure that we're not slowly poisoning our highly mobile selves. However, all types of batteries are now globally classified as poisonous waste and disposal of batteries. Therefore, battery industry has created significant efforts to recycle and reduce poisonous materials. On the other hand, Lead-acid batteries are the environmental success products. More than 97 percent of all battery lead is recycled. Compared to aluminum soft drink, beer cans, newspapers, glass bottles and tires, lead-acid batteries top the list of the most highly recycled consumer product.

Acknowledgements

The author gratefully acknowledges the guidance and discussion by Dr. Mohamed Nayel of Xi'an Jiao Tong Liverpool University for the lectures given by him.

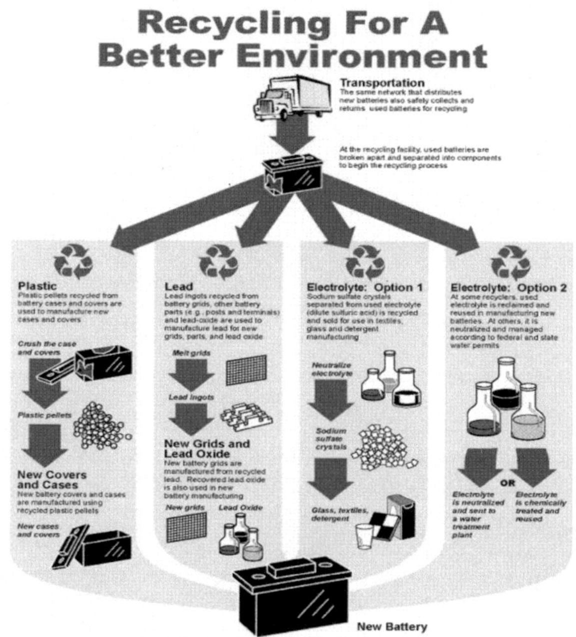

New batteries are recyclable and comprised of previously recycled materials.

(Taken from:
http://www.batterycouncil.org/LeadAcidBatteries/BatteryRecycling/tabid/71/Default.aspx)

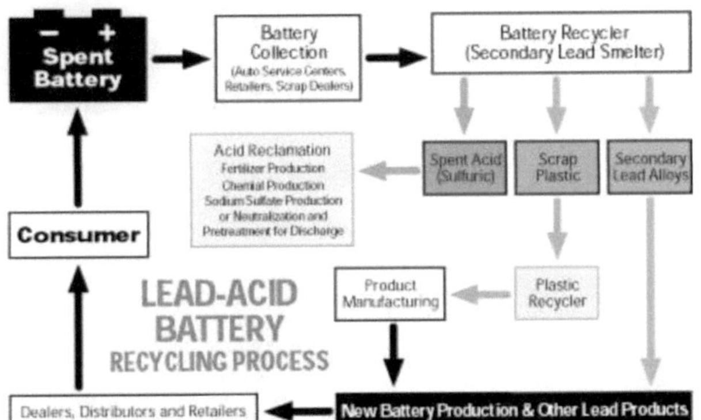

(Taken from: http://www.gravitaexim.com/Battery-Recycling/battery-recycling.html)

References

APC. (2005). APC Annual Report 2004.
http://www.apcc.com/go/2004annual/financial.cfm

Battery Council International. (2005). Battery Council International Breakdown
of U.S./North American Battery Shipments.
http://www.batterycouncil.org/BreakdownOfShipmentsChart.pdf

Battery Council International, (2009).Battery Recycling
http://www.batterycouncil.org/LeadAcidBatteries/BatteryRecycling/tabid/71/D
efault.aspx

California State University, Long Beach. (2005). Environmental Compliance.
http://daf.csulb.edu/offices/bhr/safetyrisk/envcomp.html

Earth911. (2005). Why Are Some Batteries Harmful For The Environment?
http://www.earth911.org/master.asp?s=lib&a=electronics/bat_env.asp

EERE. (2003). Energy and Environmental Profile of the U.S. Mining Industry,
Chap. 6, Lead and Zinc.
http://www.eere.energy.gov/industry/mining/pdfs/lead_zinc.pdf

Exide. (2003b). Material Safety Data Sheet. Marathon®/Sprinter®, Standard
Jar Cover
Material.
http://www.exide.com/material_safety_data/Z99MSDSMARSPR_Rev_AC_2003
0917_Marathon_and_Sprinter_Sealed_Lead_Acid_Battery.pdf

Frost and Sullivan. (2004a). World Stationary Lead Acid Battery Markets.
Research Overview.
http://www.frost.com/prod/servlet/reportoverview.pag?repid=A71401000000
&title=Research+Overview

Frost and Sullivan. (2004b). Advanced Energy Storage Technologies.
Research Overview.
http://www.frost.com/prod/servlet/reportoverview.pag?repid=D28901000000
&title=Research+Overview

Imperial College Consultants Ltd (2001). "LEAD: the facts".
http://www.ldaint.org/factbook/factbookch1.htm

Lead Acid Battery Management(2010).
http://www.sprep.org/solid_waste/documents/Solid%20Waste/Guidelines/Bat
tery%20Management%20.pdf

Leadpoison.net. (2002). LeadSafe Guide for Health Professionals.
http://www.leadpoison.net/prevent/guideprofessionals.htm

National Institute for Occupational Health and Safety. (2002) Protecting Workers' Families AResearch Agenda.
http://www.cdc.gov/niosh/docs/2002113/2002113.html

Nayel, M. (2010), "Sustainability in Practices and product development", EE12012 Sustainability Seminar.

U.S. Geological Survey. (2005a). Lead Statistics and Information.
http://minerals.usgs.gov/minerals/pubs/commodity/lead/

Venture Development Corporation. (2004). 2003 Power Protection Market IntelligenceProgram.
http://www.vdccorp.com/power/white/04/04ups.pdf